MATH NOTEBOOK

This notebook belongs to:

_____

$$5 \times$$
$$5$$
$$22 \overline{5}$$
$$5$$
$$\times \quad 5$$
$$\overline{125}$$

$$\begin{array}{r} d \\ 3 \\ \times \\ \overline{27} \\ 2 \end{array}$$

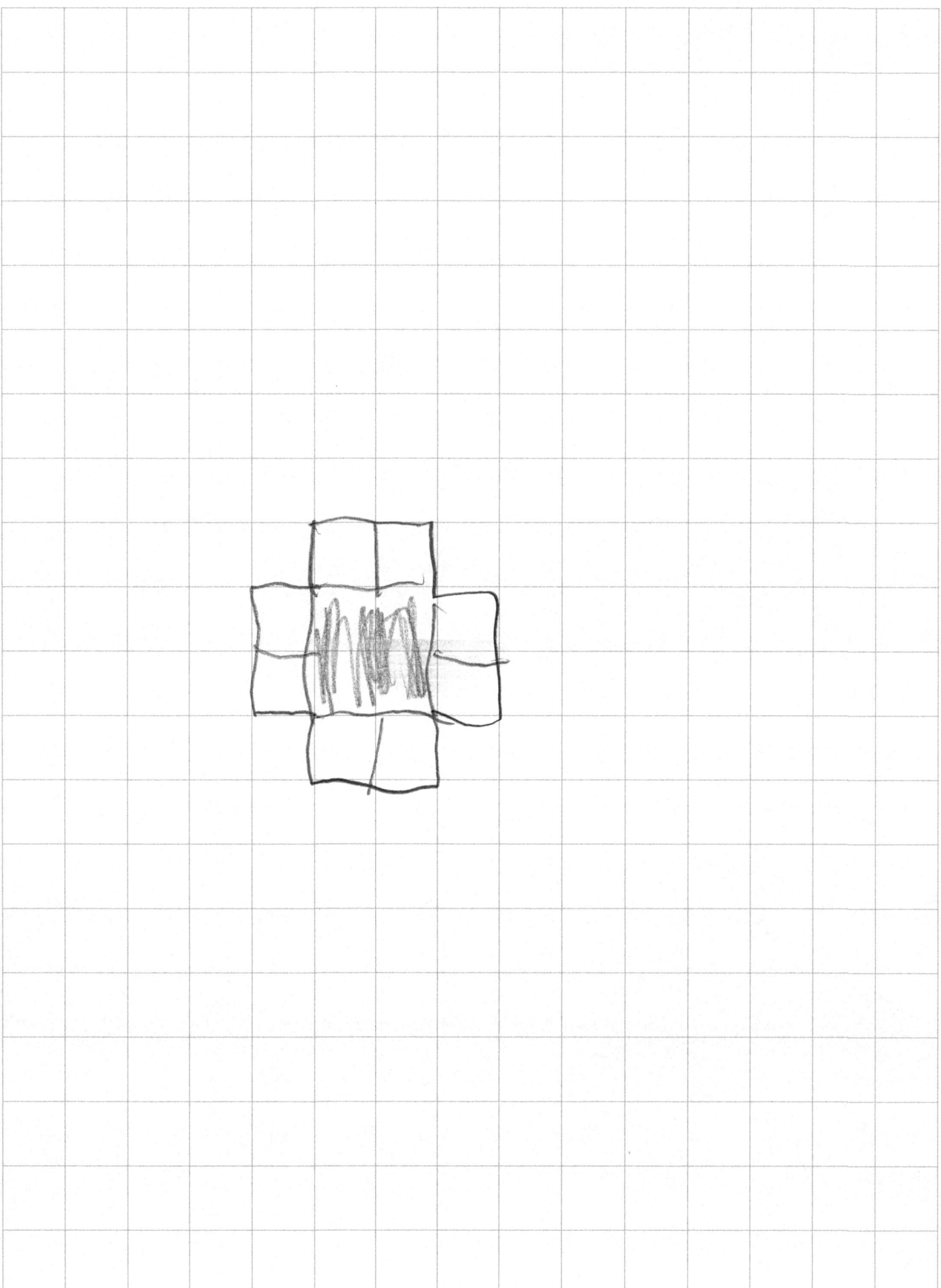

Made in the USA
Middletown, DE
16 October 2020